Men of the 1st Middlesex Victoria and St George's Rifles on parade at Winchester, c.1897. In 1894 construction of the Peninsular Barracks commenced at Winchester to replace the old King's House Barracks, which had been gutted by fire. The earlier barracks, part of which is shown, started life as a palace for Charles II, designed by Christopher Wren. The building was never completed and the shell was adapted to house French prisoners, being converted to military barracks in the late 1790s. After the fire, the remaining structure was pulled down but much of the seventeenth-century stonework was incorporated in the new buildings. Despite their imposing frontage, however, behind the façade the barracks were of a standard dormitory type used for depots.

Military Barracks

Trevor May

A Shire book

Contents

Cover: *The Royal Horse Artillery on stable parade in 1888. Detail from a painting by G. D. Giles.*

ACKNOWLEDGEMENTS

Illustrations are acknowledged as follows: Black Watch Regimental Museum, page 32 (top); Bourne Hall Museum, page 37 (top); Crown Copyright, Ministry of Defence, pages 10 (both), 14, 18 (centre and bottom), 37 (bottom); Courtesy of the Director, National Army Museum, London, page 17 (top); Copyright, Royal Commission on the Ancient and Historical Monuments of Scotland, page 7 (both); Royal Marines Museum, page 33 (top). Most other illustrations are from the author's collection.

British Library Cataloguing in Publication Data: May, Trevor. Military Barracks. – (A Shire book; 358) 1. Great Britain. Army – Barracks and quarters 2. Great Britain. Royal Navy – Barracks and quarters. 3. Great Britain. Army – History 4. Great Britain. Army – Military life I. Title 355.7'1'0941. ISBN 0 7478 0489 3.

Published in 2002 by Shire Publications Ltd, Cromwell House, Church Street, Princes Risborough, Buckinghamshire HP27 9AA, UK. (Website: www.shirebooks.co.uk)
Copyright © 2002 by Trevor May. First published 2002. Shire Album 358. ISBN 0 7478 0489 3.
Trevor May is hereby identified as the author of this work in accordance with Section 77 of the Copyright, Designs and Patents Act 1988.

Printed in Great Britain by CIT Printing Services Ltd, Press Buildings, Merlins Bridge, Haverfordwest, Pembrokeshire SA61 1XF.

William Hogarth painted 'The March to Finchley' in 1749, and it was published as an engraving in the following year. It depicts the Foot Guards being marched from London to Finchley, where George II had ordered a camp to be established as a line of defence against the Jacobite invasion from Scotland. This detail captures well the lack of discipline of the soldiers, who have as yet got no further than Tottenham Court Road. It was the desire to segregate soldiers from the temptations of civilian life that prompted many military authorities to house them in barracks rather than in alehouses and other billets.

Introduction

It was the Romans who brought barracks to Britain. They built them because they needed to. Native peoples had to be subdued, and fortified strongholds were necessary both to house the legionaries and to symbolise imperial power. The fort at Housesteads, on Hadrian's Wall, is one of the most impressive, and its barracks, storehouses and workshops accommodated 1000 men. However, the history of barrack building in Britain, both in terms of location and date of construction, shows great variation, depending on the directions from which threats were perceived and the extent of the threats.

The term 'barrack' derives from the kind of temporary hut constructed by soldiers on campaign, known in Spanish as a *barraca* and in French as a *baraque*. The first recorded use of the term in English goes back no further than 1670 and referred to the 'Irish Barracks' recently constructed within the Tower of London.

Strong constitutional objections had to be overcome before barracks became commonplace. Ever since Cromwell's New Model Army and James II's use of military power to intimidate opponents there was considerable antipathy to standing armies. The Constitutional Settlement of 1688 secured a compromise. A permanent army subject to the sovereignty of Parliament was to be maintained but was not to be housed in permanent quarters. That, it was argued, would separate the Army from the people and increase the risk that it would be used by ambitious and unscrupulous commanders to their own ends. Instead, the Army

A detachment of the Sealed Knot performs seventeenth-century pike drill at Tilbury Fort, Essex. This was designed by Sir Bernard Gomme, Chief Engineer and Surveyor General to the Ordnance between 1682 and 1685, whose design drew heavily upon the ideas of continental military engineers. By 1679 the barracks were capable of housing 220 soldiers. Only the foundations of the original brick barrack houses now remain, the buildings shown here being of a later date.

Below left: *Men of the 1st Suffolk Regiment parade at the Tower of London in the absence of the Household Brigade, whose duty this normally was. Troops had been accommodated at the Tower for centuries, and new buildings were from time to time erected to house them. The Waterloo Barracks, seen to the right of the White Tower, were completed in 1854 to a design that was intended to blend in with the historic castle.*

Below right: *The main guard of The Welsh Regiment turns out at the gateway of Plymouth Citadel, whose seventeenth-century buildings were amongst the first purpose-built housing for troops. The gatehouse represents one of the finest examples of military architecture of that period. When this photograph was published in 1896, it was feared that the gatehouse's days were numbered, but it went on to survive the bombing of the Second World War and is still the main entrance to the military establishment located there.*

would be spread around Britain and quartered amongst the general population. Billeting troops in this manner was highly controversial and had been one of the complaints against Charles I. The practice was accompanied by great inconvenience, but it was held that, paradoxically, its troublesome nature would keep the Army before the eyes of the public, who would press for a diminution, rather than an extension, of the armed forces of the Crown.

The Mutiny Act of 1689 made it illegal to quarter a soldier upon a private citizen in England (although in Scotland this practice remained lawful until 1858). Innkeepers, however, could be compelled by the terms of their licence to provide billets. This lack of permanent barrack accommodation resulted in the Army being scattered throughout Britain in the first half of the eighteenth century, often in very small detachments.

The public attitude to barracks reflects the fact that these buildings are more than mere housing for soldiers. They are strongholds (against domestic as well as foreign threats to the state). Indeed, at times it can be difficult to distinguish between a fort and barracks, as in the case of the Martello towers of the Napoleonic War period and the series of forts planned by Lord Palmerston to counter a suspected French threat in the 1850s and 1860s. Barracks are also recruiting and training establishments; and they have a role in gathering forces together before forwarding them to the seat of war. All of these functions will be explored in the following pages, with the greatest emphasis being on the barrack life of the Victorian and Edwardian armies.

The spread of barracks in the eighteenth and early nineteenth centuries

The spread of barracks in the eighteenth century was in response to threat, which might be internal as well as external.

The main overseas foe was France and, at the height of the threat, this led to a concentration of troops near the south coast, both in defence and as a preliminary to transport abroad.

But there was also a perceived menace to internal security, coming from the north of Britain rather than the south. In the industrial regions of the Midlands and the north of England, civil unrest was endemic. Across the border, in Scotland, there was a very real threat of Jacobite rebellion, at least until mid-century, while the fear of unrest in Ireland meant that many British troops had to be stationed there and accommodated in secure barracks. Edinburgh Castle was the main garrison for Scotland, and both there and at Stirling Castle ancient great halls were altered to provide barrack accommodation.

Shortly after the Jacobite rising of 1715, extensive barrack building commenced at Berwick-upon-Tweed, where the billeting of soldiers had previously caused great friction. Buildings to house 600 men and thirty-six officers were constructed as an enclosed unit, separated from the town. It was in the Highlands, however, that military posts were most necessary, and a number of defensible barracks were built, of which the best preserved is Ruthven Barracks in Badenoch, Inverness-shire. These were completed in 1724 on a site once occupied by a medieval castle. A garrison of 120 men was to be housed there, in buildings that were constructed with loopholes for musketry fire. Ruthven formed the junction of three new military roads built by General Wade. In August 1745, a sergeant and twelve men successfully repelled a force of some 200 Jacobites; but when the retreating Jacobite army

returned in February 1746 accompanied by artillery the defending force was obliged to surrender. When the Jacobites evacuated the barracks after the defeat at Culloden, they put them to the torch and they were never rebuilt. The ruins, however, remain impressive.

Ruthven was not the only fort to fall to the army of Prince Charles Edward Stuart, and with the collapse of the '45 Rebellion the government determined to construct a massive and impregnable fortress in the Highlands. The site chosen was a narrow spit of land on the Moray Firth at Ardersier, north-east of Inverness. The construction of Fort George commenced in 1753. It took more than twenty years to complete, at the huge cost of £200,000 (around £1 billion at today's prices). The fort occupied a 42 acre (17 hectare) site and was designed to

The archway entrance to the barracks at Fort George. The barracks were divided into sections like separate terraced houses, and originally eight men shared a small square room, sleeping two to a bed.

accommodate two whole infantry battalions of 1600 men. By the time it was completed the original threat had passed, despite which the establishment has continued to house units of the Army to the present day.

The location of Fort George on the Moray Firth put paid to any chance of renewed French assistance to Jacobitism, should such a movement revive. For a large part of the eighteenth century Britain was at war with France, and this conflict was to have considerable repercussion on barrack building.

Above and below: Ruthven Barracks were illustrated in Curdiner's 'Remarkable Ruins' in 1795, within seventy years of their construction. The ruins still stand, a stark reminder of the suppression of Scottish Jacobitism by the army of George II.

The Higher Barracks in Exeter were built in 1794 to house 180 cavalrymen. Their position on rising ground close to the city centre made the site a prime one for developers, and when the Army moved out the buildings were converted to private housing.

Between 1789 and 1814, the Army grew from 40,000 to 225,000 men (many of whom were stationed abroad), while the militia (a local force for national defence) was embodied in 1793 and numbered 100,000 by 1797. In 1805 it was claimed that there were 810,000 men under arms in the United Kingdom.

There had been invasion scares before, but the huge size of the French army gave rise to the fear that an invasion might be mounted at any time of the year rather than in the summer months, which had been the expectation in earlier times. As a consequence, large numbers of troops had to be maintained in the coastal counties throughout the year. And that meant they had to be housed. Yet in 1792 the whole of the accommodation provided directly by the Board of Ordnance in Great Britain and the Channel Islands was sufficient only for 20,000 or so men, distributed in forty-three fortresses and garrisons. There was no way in which the billeting system could cope with such numbers, and in many parts of the country innkeepers petitioned for relief. In June 1793 the government responded by setting up a new Barrack Department. This became responsible for all military accommodation other than in fortified places where there were artillery defences, which remained with the Ordnance Board.

The Barrack Department initially took a long-term view of national defence, and some impressive brick and masonry buildings were designed. But as the threat of invasion increased, speed of construction became of the essence, and less durable barracks were put up, some being no more than timber huts or even sod-walled cabins – a revival of the original *barracas*. Top priority was given to a chain of cavalry barracks around the coast, with a concentration in East Anglia and the southern counties. These ranged in size from stations for single troops of fifty-eight mounted men and their officers, to four-troop stations, which were strung at intervals along the chain.

As well as constructing barracks, the Department looked for suitable buildings to lease, and it acquired warehouses, factories and barns for conversion to military use. In Bristol, a lease was even taken on the newly completed (but unoccupied) Royal York Crescent, but it was never used. By 1796 the Barrack Department was able to report that forty-two permanent barracks had been built, with a capacity of 16,311. Of these, thirty-two were cavalry stations.

Vast hutted encampments were also established at such places as Glasgow, Sunderland, Colchester, Chelmsford and Portchester. Much of the work of erecting them went to large contractors. Alexander Copland, for example, had his own brickworks and sawpits and was able to erect infantry hutments for 2400 men at Chelmsford in five weeks during 1796.

The Ordnance Board was also busy, housing gunners. In the 1770s it had commenced construction of massive, permanent barracks on Woolwich Common, overlooking the dockyard and the arsenal. The façade, over 1000 feet (304.8 metres) in length, made this probably the longest residential building in Georgian Britain, and it was described by the architectural historian Nikolaus Pevsner as being comparable in scale 'only to St Petersburg'. All the ancillary buildings were at the rear, arranged on a grid pattern reminiscent of a Roman fort. Here, when the barracks were occupied in 1808, was all that was needed to support more than 3500 men and 1700 horses.

The Ordnance Board built other barracks for field and horse artillery, including Brompton Barracks at Chatham (later the headquarters of the Royal Engineers) and Wyvern Barracks, Exeter. Much of the Board's work, however, was in the construction and improvement of coastal defences, an activity that blurs the line between barracks and forts. Martello towers, for example, of which 121 were constructed along the coastlines of Britain and Ireland between 1804 and 1812, each had to house a garrison of around twenty-four gunners and one officer.

If fear of the French encouraged barrack building in the southern part of the country, then fear of industrial and political unrest pulled troops to the north, especially at the end of the

Troops depart from Euston station on the London & Birmingham Railway to put down industrial unrest in Manchester in 1842. The spread of railways greatly facilitated troop movements and avoided the necessity of billeting troops on the march. The strategic value of railways was speedily appreciated, although canals had also been used for military purposes. A great ordnance depot was constructed on the Grand Union Canal at Weedon in Northamptonshire, midway between Birmingham and London. Later, many of the warehouses were converted to barracks.

Fulwood Barracks at Preston, Lancashire, were completed in 1848 and had provision for over 1200 men, comprising a regiment of infantry, two troops of cavalry and forty-six artillerymen. The central entrance led to the infantry parade ground (shown here) and contained the chapel as well as the barrack master's quarters. There was a separate parade ground for the cavalry.

Built at a cost of just under £138,000, Fulwood Barracks at Preston still serve their original purpose, although they no longer need to awe the local population. They are the only barracks built to maintain order in the industrial north of England to survive.

Napoleonic Wars when Britain underwent a period of considerable domestic turbulence. Even in 1812, for example, when the Peninsular War was at its height, 12,000 men were stationed between Leicester and York in order to suppress the machine-breaking Luddites. Huddersfield had no fewer than 1000 soldiers billeted in thirty or so public houses in September of that year. The fear that the men would become too intimate with the people they were expected to control encouraged the authorities to believe that a measure of segregation could be achieved by the construction of barracks.

In 1840 Sir Charles Napier, Commanding Officer of the Northern District, was commissioned to write a report on the security situation in his area. The principal concern that he expressed was the need to avoid dispersing troops in small detachments, where they might become isolated. The coming of the railway helped to do away with the necessity of spreading forces thinly, for they could now be dispatched at speed from barracks located near railway stations. He therefore proposed that three new barracks be built at Bury, Ashton-under-Lyne and Blackburn (for which Preston was later substituted). These were constructed between 1842 and 1848, while older barracks at Sheffield were rebuilt between 1847 and 1854. These barracks were made defensible (at least against lightly armed crowds) by an encircling wall – although walls were often meant to keep impressionable soldiers in rather than keep a hostile populace out.

Popular disturbances diminished after the 1840s, but the ability to use troops for internal security still remained important. Indeed, it continued to be the principal strategic priority of the Secretary of State for War as late as 1888, and troops were called out 'in aid of the civil power' on twenty-four occasions between 1869 and 1910.

The Westminster Guards Barracks (later known as Wellington Barracks) were designed by Sir Frederick Smith of the Royal Engineers and Philip Hardwick, a civilian architect best known for his work at Euston station. Opened in 1834, their classical style aptly fitted the location alongside St James's Park and close to Buckingham Palace. This engraving of 1867 shows constables of the Metropolitan Police practising cutlass drill at the time of the Fenian outrages. The establishment of the Metropolitan Police in 1829, and of county police forces by the 1850s, gradually diminished the role of the Army in civil policing, although it remained a last resort and barracks were located accordingly.

Barrack blocks in the Red Fort, Delhi. After the Indian Mutiny the Delhi garrison was strengthened, and many ancient buildings within the fort were demolished to make way for these new barracks. The galleries were intended to provide some protection from the heat. Similar galleried buildings were sometimes erected in England, but the aim there was rather to provide a covered space for drill in wet weather.

Victorian barracks

It was factors other than a general growth in the size of the Army that influenced the Victorian concern for barracks. There were around 204,000 men serving in the Army at the end of the Napoleonic Wars in 1815, but thereafter numbers fell rapidly and were not to reach a similar height until the Crimean War, forty years later. There were over 223,000 men under arms in 1855, but once again numbers fell with the resumption of peace. The South African War saw an army of 430,000 in 1900, while it numbered 733,000 in 1914 and nearly 3,760,000 in 1918. By no means all of those men were to be found at home, for in the latter half of the nineteenth century around half of the Army was stationed in posts spread throughout the Empire, and many regiments spent well over 60 per cent of their time abroad.

Hostilities gave rise to a need for increased accommodation, making the Crimean War and its aftermath a significant period for barrack construction, while even an imagined threat – such as that felt from France in the late 1850s and the 1860s – left its mark. Queen Victoria attended the inauguration of the improved French naval base of Cherbourg in 1856 but her Prime Minister, Lord Palmerston, took a more sober view. His fear of a possible invasion was increased in 1860 when the iron-clad warship *La Gloire* was launched, and Palmerston set up a Royal Commission, which recommended the construction of nineteen new forts and fifty-seven batteries around the coast. Yet the French were Britain's allies in the Crimean War against Russia, a war that is often credited with shaking up the British Army – although perhaps of equal importance in this matter was the death of the Duke of Wellington in 1852, which removed a powerful restraining hand upon reform.

The Methodist garrison church at the Trimulgherry cantonment, Secunderabad, India. In the days of the Empire, many soldiers spent much of their time abroad. The general antipathy towards church parades may have been tempered a little by chapels such as this, which, architecturally at least, was a reminder of home.

Crownhill Fort formed part of Palmerston's defensive ring against a French invasion and is now one of only two examples surviving in their entirety. Designed by Edmund du Cane (better known for his designs for Wormwood Scrubs prison), the fort was completed in 1872 to defend the Royal Dockyard, Devonport, from a land attack. It was never used in earnest, and by the time it was finished Napoleon III (from whom the threat was perceived) was in exile in England, where he died the following year. The casemated 'officers' quarters' also included additional accommodation for up to thirty-five other ranks. A restored barrack room gives a somewhat sanitised impression of living conditions in the 1890s, although the lack of space between beds is realistic enough.

13

The south block of the Royal School of Military Engineering, Brompton Barracks, Chatham. In 1812 the Royal Engineers took over what had previously been an artillery barracks constructed by the Ordnance Board and completed in 1806. In 1822 the Royal Engineers assumed responsibility for barrack design. A considerable body of experience was built up, and officers from the Corps were sent out to design and construct barracks throughout the Empire.

In 1855 the Ordnance Board was abolished and, although the Royal Engineers remained responsible for barracks, that responsibility was now exercised under an Inspector General of Fortifications at the War Office. In the same year, a Barracks Accommodation Report was presented to Parliament. One of the most significant aspects of the report was the importance placed upon improving the character of the soldier by providing healthier living conditions: better sanitation; improved arrangements for cooking and eating; and better provision of recreational and educational facilities. There was certainly room for improvement in each of these fields.

It was said that in the 1850s there were only seven barracks outside London that were capable of housing 1000 men. The dispersal of the Army throughout the country in small units may have met the need for a force to aid the civil power at times of civil disturbance, but it made difficult the holding of training exercises or manœuvres on any large scale.

In 1852 the Army held training exercises at Chobham, in Surrey. These proved so successful that soon afterwards 10,000 acres (4000 hectares) of sandy heathland were purchased at Aldershot for the establishment of a concentration area and advanced training camp, close to both London and the Channel ports. Work began in 1854, and by 1856 wood and brick huts for 20,000 men had been erected, each hut holding twenty-two men. Permanent barracks were erected between 1854 and 1859.

Aldershot went on to become almost the home of the British Army, while another garrison town, Colchester, had similar origins. There had been a garrison in this ancient Essex town

The soldiers' reading room at St Mary's Barracks, Chatham, in 1856. As the effective range of artillery increased in the eighteenth century, vaulted casemates became commonplace as a protection against bombardment. To assist in defending the dockyards at Chatham, the Ordnance built St Mary's Barracks, which consisted entirely of casemates, with a capacity of over 1100 men. Construction took place between 1807 and 1812, when the barracks were given over to French prisoners of war. Ventilation was always a problem with casemates, which makes it somewhat surprising that the barracks were reopened in 1844 specifically to house time-expired men from the Empire, or those invalided home because of their health. The accompanying text describes the situation of the barracks as 'remarkable for salubrity of air'. Quite clearly, very little of it found its way inside the casemates themselves.

The original wooden huts at Aldershot were gradually replaced by permanent buildings. These brick 'bungalow' barrack blocks were built in 1894. The last surviving pair, they now house the Aldershot Military Museum.

The riding school of the Beaumont Cavalry Barracks at Aldershot was built between 1854 and 1859. Its clear span of 180 by 60 feet (54.9 by 18.3 metres) made it one of the finest riding schools in Britain when it was opened. Winston Churchill was but one of thousands who did their military-riding training here.

Colchester Camp, from the 'Illustrated London News', 19th January 1869. Built on large green field sites, hutted barracks such as Colchester and Aldershot had the merits of ample light and air. The inevitable forces of economy, however, meant that many of these huts remained in use long after they should have been replaced, and they constituted a considerable fire risk. Those at Colchester remained until the end of the nineteenth century, while the temporary wooden chapel of 1856 remains in use as the garrison church to this day.

Officers of the 86th Regiment (later The Royal Irish Rifles) relax at the Curragh Camp, County Kildare, c.1860. Construction of the camp began on 18th March 1855, and by 9th July accommodation for 5000 men had been completed. Two civilian building contractors employed between them 1100 men on site, while construction of the buildings for men and non-commissioned officers (together with stables) swallowed up 1,100,000 tons of bricks, 3800 tons of timber and 5 tons of nails.

for centuries, and a temporary hutted camp had been constructed there between 1794 and 1799. Permanent barracks were also built, but these were demolished in 1816. In January 1856 a new hutted camp was opened along the lines of Aldershot. At the end of the Crimean War, instead of being allowed to run down, the camp was placed on a permanent footing, and by 1864 it had doubled in size through the completion of new cavalry barracks.

Portman Street Barracks, in London's West End, were built in the eighteenth century for cavalrymen but in the 1850s were used to house about 500 Foot Guards. In 1859 it was observed that when the room illustrated was occupied by its twenty-two inhabitants 'it requires strong nerves on the part of one who would put his nose into it'. Bad ventilation was the curse of many barrack rooms, making them nauseous places.

An army camp, to defend the western approach to London, existed at Hounslow from the seventeenth century, and in the eighteenth century permanent barracks were built there. In 1860 designs were produced for married quarters to house forty-two families, each living in a one-roomed apartment containing a bed, a child's crib, two cupboards and a fire for cooking. Lavatories were located at either end of the verandas (below). Under Cardwell's localisation programme, Hounslow became the depot for both the Middlesex and the Metropolitan Districts. Further building works were called for, and Hounslow became one of the largest local depots in Britain. The Hardinge Block (above) was constructed to house cavalrymen who, for reasons of health, were boarded away from their horses rather than above them, which had been the common practice. This building, while austere in design, nevertheless represents a marked improvement in the living conditions of the soldier.

The Victoria Barracks at Bodmin were commenced in 1859, making much use of local granite. The majority of the buildings are now demolished, but one imposing block remains, behind a perimeter wall containing rifle slits. The barracks were built to house the Duke of Cornwall's Militia at a time of renewed fear of French invasion. It is therefore ironic, as the architectural historian James Douet points out, that the chosen style was that of 'a minor Loire château'.

Renewed fear of a French invasion in the 1850s led to a revival of the militia, and in 1859 a Volunteer Force was established. As service was part-time, the required accommodation consisted of secure armouries and facilities for drill, rather than housing. The new Blackheath drill shed for the 25th Kent Rifles was built at the expense of its commanding officer, Captain Rucker. The drill room itself was 100 feet (30.5 metres) long and 53 feet (16.1 metres) wide. (Engraving from the 'Illustrated London News', 15th February 1862)

The Crimean War brought fresh thinking to the question of barrack design and construction. There was a growing appreciation of the importance of morale to the efficiency of a fighting force, and a realisation that this depended to a large extent on the conditions to which the soldier was subjected, including his accommodation. A Royal Commission on the sanitary condition of barracks reported in 1861 that whereas the mortality rate of male civilians aged twenty to forty was 9.8 per thousand, that of soldiers in barracks was 17.11 per thousand – despite the fact that the unhealthiest were filtered out at the time of recruitment. The Royal Commission inspected 162 barracks and calculated that (on the space allocated to convicts in prison) barracks were overcrowded by more than one third. The Royal Commission made a number of recommendations, which were accepted by Parliament. Amongst other provisions, new minimum space requirements were to be adopted (17 cubic metres of air space and 5.6 square metres of floor space per man); bath-houses and privies were to be improved; recreational facilities were to be provided; and a more structured exercise regime was to be introduced, using purpose-built gymnasia. A start was made on improving the nation's barracks, and some of the worst examples were closed; but progress was both slow and patchy.

Recruiting sergeants at St George's Barracks, Charing Cross, London, in 1896. Although this was technically the depot of the London Recruiting District, about one quarter of all recruits for the Army were enlisted here, for many men preferred to come to London to enlist rather than join up at their local depot. Apart from recruits to the Royal Engineers, the Army Service Corps and the Medical Staff Corps, no character references were required provided that a potential recruit had 'nothing of a suspicious nature about him'.

The sweeping reform of the Army instigated by Edward Cardwell, Secretary of State for War between 1868 and 1874, introduced a new twist. The Prussian Army enjoyed spectacular military successes in the struggle for German unification between 1864 and 1871, with conscripts drafted into regiments organised on a local basis. Reformers in Britain took up the idea, seeing it as a means of encouraging recruitment, improving training and linking regular units with those of the militia. Cardwell therefore proposed that Britain and Ireland be divided into sixty-six infantry districts, twelve artillery districts and two cavalry districts. Each district would have a regimental depot, which, it was hoped, would appeal to the local patriotism of recruits, and where they might undergo their basic training. Over half the towns chosen for such regimental depots already had regular army or militia barracks that could be adapted, but elsewhere quite new facilities were required. A costly building programme therefore commenced, resulting in twenty-two new depots being built by 1880. The 'localisation programme', as it was called, led to the provision of 13,350 beds for single men, 2014 quarters for married soldiers and accommodation for 542 officers. In addition, 977 hospital beds were provided, and stabling was constructed for 1118 horses. Poor living conditions survived in many places, especially where 'temporary' wooden huts from the Crimean period still continued in use, but the sanitary improvement of British barracks is evident from the fact that by 1897 the death rate amongst the home Army had fallen to 3.42 per thousand.

Life in nineteenth-century barracks

The life of the soldier in barracks could be dull, as the *Quarterly Review* observed in 1859:

> Perhaps . . . no living individual suffers more than the soldier from *ennui*. He has no employment save the drill and its duties; these are of a most monotonous and uninteresting description, so much so that you cannot increase their amount without wearying and disgusting him. All he has to do is under restraint: he is not like a working man or an artisan; a working man will dig, and his mind is his own; an artisan is interested in the work on which he is engaged: but a soldier has to give you all his attention, and he has nothing to show for the work done. He gets up at six; there is no drill before he makes up his bed and cleans up his things: he gets his breakfast at seven; he turns out for drill at half-past seven or eight; his drill may last half an hour. If it be guard-day there is no drill except for defaulters. The men for duty are paraded at ten o'clock; that finishes his day-drill altogether. There is evening parade, which takes half an hour, and then his time is his own until tattoo, which is at nine in winter and ten in summer. That is the day of a soldier not on guard or not belonging to a company which is out for Minié [rifle] practice.

The imposing guardhouse (right) of the Peninsular Barracks, Winchester, with its Roman Doric colonnade. The entrance to the Portman Street Barracks in London was less grand, and George Godwin observed that 'But for the small boys and patient maidens loitering round the entrance . . . it might be supposed to lead to stables or a builder's yard'. The engraving of the Portman Street guardroom at night (below) comes from Godwin's 'Town Swamps and Social Bridges', published in 1859, in which he wrote scathingly about the conditions in London barracks. When on guard, men were required to be fully dressed the whole time. The only possibility of rest was on sloping so-called beds with a raised wooden pillow.

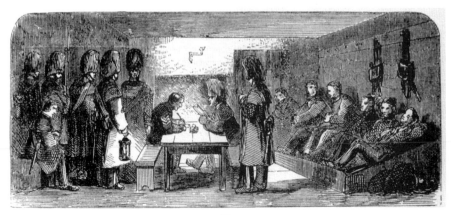

A soldier of The Royal Horse Guards is tried by a barrack-room 'court martial'. This carefully posed photograph of 1896 depicts the event as a light-hearted comedy, but, in reality, such rough justice would have been no joke to the soldier who provoked the censure of his comrades.

The position had hardly changed by 1900, when Captain W. E. Cairnes wrote:

Those hours of the morning which the soldier does not spend on parade he usually occupies in fatigue, sweeping up the flotsam and jetsam of a barracks, scrubbing the floors of the officers' or sergeants' mess, cleaning out the canteen or recreation room, or cleaning roads or paths about the barracks from the encroaching grass. Neither the parades, the guards, nor the fatigues which occupy his time do anything whatever to improve his military efficiency, yet in this dreary routine the soldier spends considerably more than half of every year which he passes within the United Kingdom.

J. E. A. Troyte, a 'gentleman-ranker', described a typical day in September 1873:

Turn out of bed at 6, and, after a short wash and making up bed, parade, 6.30–7.30; breakfast at 7.45; then sit round table and peel potatoes, etc., for dinner; time for general polishing up of person and clothes; parade, 9.30 – for an hour; and for recruits another parade about 11.30; dinner, 12.45; parade, 2.30; tea, 4.30; then a free time until 9.30, when the roll of each company is called, the orders for next day read out, men warned for any special duty (such as guard), and to bed, and 'lights out', at 10.15.

From 'reveille' to 'lights out', the day was

Orderly officers receive meat rations, c.1900. The daily allowance of meat per man was 12 ounces, weighed with the bone. For most of the nineteenth century, cooking facilities were limited to a large copper boiler. After boiling, and with the bone removed, as little as 7 ounces of tough, stringy meat per man was left.

The regimental kitchen of the 3rd Dragoon Guards at Woolwich was a stark affair in 1896. Although it was in the charge of a cook-sergeant, trained at the Aldershot School of Cookery, most of the cooking was done by two men selected from each of the four squadrons of the regiment.

punctuated by bugle calls, to which the men added words as an aid to memory:

I called 'em, I called 'em
They wouldn't come, they wouldn't come
I called 'em, I called 'em
They wouldn't come at all.

So ran the bugle call for 'rations' – the rations themselves remaining largely unchanged from 1813 to the end of the century. Each man was entitled to a daily ration of three quarters of a pound of meat and one pound of bread. The monotony of the basic diet of meat, bread and potatoes was exacerbated by the limited cooking arrangements for most of the century, which meant that boiling was the only culinary option. From 1857 an extra allowance was made for the purchase of vegetables and condiments, but it was claimed in 1876 that army food was as coarse, tasteless and monotonous as that supplied to convicts in prison.

It might take an infantryman several hours to clean all his equipment and, when done, it had to be guarded from dust or from spotting by any liquid. A drop of oil on a pipeclayed belt was hard to remove, while a sharp shower could wash pipeclay from the soldier's belts on to his uniform. Proprietary pastes such as 'Blanco' claimed to make cleaning easier, but military authorities held out against more easily cleaned brown leather straps, claiming that they could not be kept to precisely the same shade.

Coldstream Guardsmen under instruction in volley firing on the barrack square. Before practising at the butts, soldiers learned the rudiments of musketry in barracks. There was much criticism of the small quantity of ammunition issued for training, and inadequacies in musketry led to changes being instituted during the South African War. Even so, in 1901 it was claimed that 'It is good for the soldier if his captain can say of him, "He is a good shot" but it is better he can say, "He is a clean soldier". In itself this is not a bad hint at the real defects of our military system.'

Soldiers responded in various ways. They assuaged the pangs of hunger by smoking and drinking, or they purchased additional food from their pay, often eating outside the barracks. Otherwise, individual tastes were not catered for. Cairnes wrote:

> The orderly man clatters in at the door with a steaming can of tea, from which he rapidly fills the basins, the milk and sugar having been already added before the tea left the cookhouse. No allowance is made in barracks for men of fastidious tastes; if a man prefers his tea unsweetened he can go elsewhere; the taste of the majority is alone consulted.

And the basin itself was a multi-purpose utensil. Sir William Robertson, who enlisted in the 16th Lancers in 1877 and rose to the rank of Field Marshal, observed that 'the basin [is] used in turn as a coffee cup, tea cup, beer mug, soup plate, shaving mug, and receptacle for pipe-clay with which to clean gloves and belts'.

Such was the military passion for a smart turn-out that the humble basin probably spent more time in use as a cleaning vessel than for eating and drinking. So seriously was the matter taken that the term used for cleaning and polishing kit

Two good-conduct stripes no doubt helped this Seaforth Highlander to be selected for a carefully posed mock-up photograph of a kit inspection in the 1890s. Each piece of equipment had to be placed in a particular manner and a specific place, the intention being to make it easier for an inspecting officer to spot missing items for which the soldier would be held responsible. Horace Wyndham wrote in 1899 that 'Occasionally a recruit is . . . detected in such depravity as that displayed by placing his fork on the wrong side of his spoon, or his blacking-tin where his soap should be. When such instances occur the wrath of the outraged colour-sergeant is terrible to witness.'

Men of the 2nd Battalion Gloucestershire Regiment at bayonet drill in Raglan Barracks, Plymouth, in the mid 1890s. Raglan Barracks were built between 1854 and 1856 to designs by Captain Francis Fowke, Royal Engineers. Fowke was one of the best-known army architects, although his most famous work is civilian, namely the Albert Hall in London.

was that of 'soldiering'. Like apocryphal librarians, only happy when all the books are on the shelves, military authorities only seemed happy when the men were smartly turned out. Any old adjutant will tell you, said Captain W. E. Cairnes in 1901, 'that there is only one thing which makes troops more "unsoldier-like" than musketry or manœuvres, and that is active service'. Field training suffered in consequence. Men would not be asked to lie down on the barracks drill field if the ground was dirty or damp for fear of spoiling their uniforms.

Horace Wyndham, writing in 1899, described the efforts that were necessary before a parade:

> If greatcoats are to be taken, these have to be rolled into neat cylindrical shapes and strapped firmly to the men's belts. To prepare a coat for this purpose is an art that is only acquired by constant practice, and its exercise demands, moreover, the intelligent co-operation of four men. The first thing to do is to spread the garment at full-length on a table – the outer edges are then turned inwards, and the whole folded into a narrow compass with the collar and the bottom of the skirt inclined towards the centre. Then, when a properly rectangular figure has been obtained, two men hold the bottom corners firmly, while two others commence to roll the coat from the collar downwards. As soon as this process is completed, the ends are tightly secured with pieces of string, to prevent the roll unwinding . . .

A large part of the private soldier's day was taken up with drill, and the parade ground was a fundamental element of barrack design and layout. Drill came into new prominence in the eighteenth century when large bodies of musketmen had to be manœuvred in column or line on the battlefield. In 1792, drill regulations were drawn up, based on David Dundas's *Principles of Military Movements* (1788). Battlefield movements were complex, and it was reckoned that two years of drilling were required in order to master them all. The 1792 Regulations, revised by Major General Sir Henry Torrens in 1824, remained the basis of drill throughout the nineteenth century, although in the 1870s and 1880s 'parade drill' and 'battle drill' came to be separated. Even so, Dundas's Eighteenth Manœuvre, 'Advancing in Line', remains in use to this day. Drill was seen as preparation for battle in ways other than the manœuvring of troops, for it was regarded as essential training in discipline and obedience. There were critics of the system who argued that it turned men into dull cogs in a machine; while others held that it did little to increase fitness, for it fixed the chest in one position and

Great concern was shown at the time of the Crimean War about a perceived deterioration in the physique of recruits, which resulted in the Army giving serious consideration to gymnastic training for the first time. An Army Gymnastic Staff was formed at Aldershot in 1860, where the first army gymnasium (shown here in the 1890s) was constructed. In 1862, an order was issued that all barracks should have gymnasia. Those which survive from this period are at Brompton Barracks, Chatham, and at Sandhurst, the latter now converted to a library.

In all barracks, it was necessary to provide accommodation for the regimental tradesmen. The illustration below shows the saddle-tree maker's shop of the 3rd Dragoon Guards, who were based at Woolwich. The antelope skulls are not stock-in-trade, but are trophies shot in South Africa by two officers of the regiment! (Illustration from the 'Navy and Army Illustrated', 4th September 1896)

The officers' mess of the 1st Life Guards at Knightsbridge Barracks in the 1890s. The Guards, at least when stationed in London, differed from other regiments in that many officers lived outside barracks, which was more convenient for the active social life that most of them led. For this reason, the officers' mess came late, and the Wellington Barracks at Westminster had no mess before the First World War.

A company barrack room in the 1890s (regiment and barracks not known). One man uses his folded bed as a makeshift easy chair while reading the newspaper. Cuttings from the illustrated papers and some military prints decorate the walls, together with more personal photographs and mementoes.

only exercised the lower limbs – hence the interest in military gymnastics from the 1860s.

Non-commissioned officers, specialist tradesmen (such as armourers and saddlers) and others with special duties (for example officers' servants) were occupied for much of the day, but private soldiers had much free time. Boredom, together with uncongenial surroundings, enticed many to spend as much of their free time as possible outside the barracks. In 1860 the town of Aldershot had twenty-five public houses as well as forty-seven beer houses (generally reputed to double as brothels). At the same time, the barracks contained eighteen canteens.

Since the beginning of the nineteenth century canteens had been incorporated in barracks, but until

Amongst the many expenses to which an officer was liable was that of providing his own furniture to supplement (or replace) the government issue of a small table, two Windsor chairs, fire-irons, fender and coal scuttle. Military furniture was designed to be portable, and the cost of transferring it from one posting to another was borne largely by the officer himself. (Advertisement from the Army List of 1871)

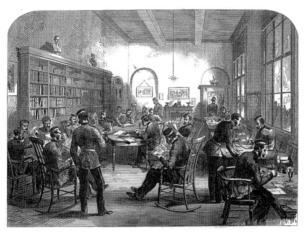

The reading room of the Guards' Institute, Vauxhall Bridge Road, from the 'Illustrated London News', June 1869. The institute was under the patronage of the officers of the whole Brigade of Guards. Recreational facilities were provided as well as some vocational training in tailoring and carpentry for men who wished it (a great asset to an ex-soldier seeking employment). Refreshments 'of a wholesome kind' were available, and these included beer and wine – the line only being drawn at 'ardent spirits'. On account of their ceremonial duties, the Guards spent long periods in London, and these regiments were at the forefront in providing welfare provisions for the men.

1863 they were run by private contractors for their own profit. At that date canteens came under regimental control and profits were ploughed back. By 1900 barrack facilities included libraries and games rooms, canteens and grocery shops. NAAFI (Navy, Army and Air Force Institutes) dates from 1921.

There can be no doubt that these new facilities improved the life of the soldier in barracks, but they did not develop without a certain resistance. Although the authorities might have wished to encourage a more productive use of leisure, the attitude of many soldiers was that, once their duty had been done, they should be left free to use their time as they wished, which often meant drinking, gambling and resorting with women.

" Comrades! Is it your ambition to figure very often in the GUARD-ROOM and CELLS? Drink is the most speedy road to those unenviable places.'

A sergeant in a Highland regiment preaches the value of temperance. 'Comrades!', he asks, 'Is it your ambition to figure very often in the GUARD-ROOM and CELLS? Drink is the most speedy road to those unenviable places.' The Army Temperance Association was formed in 1893 and had 20,000 members within a few years. From whatever cause, drunkenness certainly diminished; whereas 51,501 fines were levied in 1872, the number had fallen to 9230 by 1912–13. (Engraving from an issue of the 'British Workman', 1858, devoted almost entirely to the problem of drink in the Army)

Married quarters in a London barracks at the beginning of the twentieth century. Lit by gas, and with a good kitchen range, such a room matched, or even surpassed, in quality that enjoyed by most workers' families. But the soldier's wife was still subjected to a discipline that many civilians would have found irksome.

Women in barracks

The attitude of the Army towards marriage and towards women in barracks was ambivalent. On the whole, marriage was looked upon as a liability, resulting in lax discipline, diminished efficiency and reduced mobility. Yet, at the same time, the nineteenth century placed 'family values' at the centre of the moral code. The two viewpoints were reconciled to some degree by transferring some of the imagery of the family to the military itself, so that the Army became the soldier's 'family', and he was expected to be 'married to the regiment'. Many men found it hard to be faithful to that spouse, however, and poor barrack conditions and the desire for feminine company led many to resort to prostitutes. The result was a serious problem with venereal disease. The answer, thought many, was a more benevolent attitude towards marriage, and in the aftermath of the Crimean War a new-found appreciation of the role of morale in military success led in the same direction.

If the regiment was the family, with the colonel its paterfamilias, then there was a recognisable hierarchy amongst the female members also. Army culture led to the distinction between 'officers and their ladies; sergeants and their wives; and soldiers and their women'. The distinction went as far as washing (for women were expected to do the regimental laundry work), with officers' washing being reserved for sergeants' wives.

From 1685, permission to marry had been necessary. In the early nineteenth century the norm was six men per company given the privilege, although there was a great variation between regiments, much depending on the view of the commanding officer. The proportion was highest in the Guards regiments, partly because they rarely went abroad and were largely tied to their London base, but also because the men were considered to be of a higher calibre. Army chaplain Norman MacLeod wrote in 1863 that:

> St Paul tells us that marriage is honourable in all: but the authorities at the Horse Guards [War Office] affirm that marriage is honourable only in the case of 6 soldiers in every Company who have received

the permission of their commanding officers, and decidedly to be disapproved of and discouraged in the case of all others.

When permanent barracks came to be established at the end of the eighteenth century, the only provision that was made for married men was the 'corner system', whereby a corner of the barrack room would be curtained off for their use. No provision for extra furniture was made until 1838, when extra beds and bedding might be supplied. Inevitably, the corner

system came in for considerable public criticism on the grounds of indecency. In 1845 the *Quarterly Review* wrote of the soldier's wife under such conditions:

> What shall we say of the state of her feelings till she has become utterly hardened, while a dozen men, every night and every morning, are stripping and dressing in her presence? Or shall we ask what her husband feels when his duty comes for guard, and he is forced to leave his wife alone in such a place?

'Tommy Atkins Married – Past and Present'. From the 'Graphic' of 12th January 1884. The separate engravings cover many aspects of the life of a soldier's wife, including living conditions, regimental work, military discipline and travel at home and abroad. The central image is that of the woman 'off the strength', or married without permission. As her husband leaves for a tour of duty overseas, she and her children are left behind in England, where they must fend for themselves.

A page from the Standing Orders of the 73rd Foot, 1858. Paragraph 3 lays down that 'Non-commissioned Officers and soldiers are not to marry without the consent of the Commanding Officer. Any individual infringing this order will subject himself and his family to great misery; his wife will not be allowed in Barracks, nor have any privileges of soldiers' wives, nor be recognised in any way.' This page is from the personal copy of Lieutenant Hugh Hackett Gibsone. The pin-ups with which he has decorated it suggest that there were times when he, too, lacked female company.

In the 1850s, as part of general barrack reform, the Army authorities took a greater interest but, even in improved barracks, married couples were expected to share rooms. Official figures showed that in May 1857 there were only twenty married quarters in the 251 military stations operated by the Army, and that these provided accommodation for only 541 soldiers and their families. Aldershot, which at that time had 18,018 men in barracks, had no separate provision for married men whatsoever.

Women in barracks were expected to work. Washing for the men provided a pittance of an income, and there was also sewing to be done. Mainly, this was general repair work for the men, but in 1879 3500 soldiers' wives, widows and daughters were employed at military shirt-making on behalf of Army clothing depots, each shirt bringing an income of $8^1/_2$ pence. At Aldershot, some wives worked in government tent and bedding stores, while others sewed powder bags at the Royal Arsenal, Woolwich.

Married soldiers' quarters at Albany Street Barracks, London. The room illustrated in this engraving of 1859 was occupied by three married soldiers and was situated above the stables. The only separation from other occupants of the room consisted of curtains drawn around the bed. The accompanying text concluded that 'Anything more demoralising or destructive to the self-respect of men, women, and children, could not well be imagined – nothing can stand against its effects.'

Royal Marines at Flathouse Quay in Portsmouth Dockyard, c.1890. On the left are the accommodation hulks 'Duke of Wellington' and 'Asia'.

Naval barracks

In the heyday of sail, naval crews were raised (often by press-gang) when a ship was ready to sail and were laid off when the ship returned from its tour of duty. The need to provide shore accommodation for naval personnel therefore hardly arose. But by the middle of the nineteenth century technical advances in propulsion and naval operations made the dispersion of crews wasteful, and in 1853 the paying off of crews at the end of each commission ceased. As a result, means were needed to house crews between one posting and the next.

The clock tower of the Royal Naval Barracks, Devonport. 'Keeping one's time by the dockyard clock' could never have been easier than here. Naval conservatism is reflected in the fact that, despite the introduction of the electric telegraph, the tower was still surmounted by a semaphore.

A barrack room at the Royal Naval Barracks, Devonport, in 1897. In contrast to Army barracks of the time, the seamen's dormitories strongly reflected the gun-deck of a warship. The rolled hammocks can be seen in a rack on the right.

A room off the canteen at the Royal Naval Barracks, Devonport. Here, 'men may sit in friendly groups and discuss their beer and the affairs of the nation, if it so please them, or indulge in a smoking concert or "sing-song", as in [this] illustration, where the schoolmaster accompanies a quartet on the harmonium'.

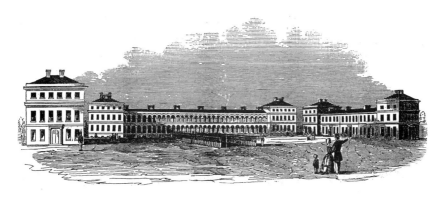

The Royal Marine Barracks at Woolwich (later Cambridge Infantry Barracks). When completed in 1848, the barracks provided accommodation for over 1000 men. A veranda, which could be used for drill in wet weather, ran the length of the building. Another feature of the building was a weight-driven Archimedean screw, which could deliver warm air from a furnace, or cool air in summer.

During the Napoleonic Wars naval hulks were used to house French prisoners; and they were used in a similar manner to house convicts awaiting transportation. Convict hulks were given up in the 1850s, but the use of hulks as depot ships for the Royal Navy continued. Not for the first time, nor the last, servicemen were housed in conditions no better than those of criminals.

It was not until 1879 that work began at Devonport on the first shore-based naval barracks, the first blocks of which were occupied ten years later. The delay was due more to naval obstruction than to slowness of construction, for the idea of housing men in 'stone ships' was repugnant to some. As the *Navy and Army Illustrated* put it in 1897, 'Naval officers are often prone to jump to the conclusion that any innovation probably involves the introduction of some hitherto unknown evil.' Originally known as Keyham Barracks, but renamed HMS *Drake* in 1934, the buildings could accommodate 4895 men when completed in 1907. At around the same time similar barracks (though built of brick rather than stone) were constructed at Portsmouth and Chatham.

While the Royal Navy was slow to provide barracks, the Royal Marines were the first complete British corps to be thus accommodated. That this should be so is no doubt due to the difficulties involved in keeping in order a force that was stationed for so long in the same towns – unlike army units, which could be moved around. Between 1765 and 1781, therefore, Royal Marines barracks were established at Portsmouth and Chatham, and also at Plymouth, where the Stonehouse Marines Barracks survive as the only English regimental barracks of the period.

Twentieth-century barracks

At the outbreak of the First World War, home barracks had room for 174,800 men, but such was the enthusiasm for volunteering that, by the end of the year, eight times that number had joined up. What was now required was not more permanent barracks, but rapidly constructed camps. Until 1915 much use was made of tents, but these were then phased out in favour of hutted camps. Many of these were wooden huts of a type that had been used for decades, but in 1916 Captain Philip Nissen introduced his design for the corrugated iron hut that was to bear his name. Catterick Camp, near Richmond in Yorkshire, had accommodation for 40,000 men by 1915, and subsequently went on to become a permanent depot.

In the First World War the need arose to house service personnel of a new breed – airmen and women. Much of the barrack construction then and between the World Wars was at Royal Air Force stations. A high regard was shown to design, and ideas were put forward by the Royal Fine Art Commission, whose members included the eminent architect Sir Edwin Lutyens.

The Royal Horse Guards' 'cooking-engine', capable of preparing meals for 1000 men, in use during the autumn manœuvres of 1893. The Army only tardily accepted the need for large-scale training exercises, but the Military Manœuvres Act of 1898 led to the purchase of 42,000 acres (about 17,000 hectares) of Salisbury Plain for that purpose. To serve the needs of the training ground, a new barracks was built at Tidworth, a village on the borders of Hampshire and Wiltshire.

The hutted barracks for the Royal Fusiliers were erected at Woodcote Park, Epsom, in 1914. At the end of the war, thousands of such huts were offered to the general public at knockdown prices and were used for a variety of purposes. The advertisement from the 'Bazaar, Exchange & Mart' of 31st May 1919 suggests to would-be purchasers of the officer's hut that 'If partitions are added [these huts] will make 2, 3, 4, or 5 rooms and form excellent dwellings'.

The situation since the Second World War has been one of slimming down. Large-scale amalgamation of regiments between 1958 and 1961, and again in 1967, together with the ending of National Service in 1960 and moves towards fully professional but smaller armed services, meant that the need for accommodation declined. Many barracks were demolished;

The barrack block at RAF Hullavington, Wiltshire, which came into service in 1937. Air Ministry buildings between the two World Wars developed a distinctive style, sometimes referred to as 'RAF Georgian'. Despite the military urgency of rearmament, great care was taken with design. The buildings at Hullavington, for example, were faced in Cotswold stone in response to prompting from the Council for the Preservation of Rural England.

others found a new lease of life filling a different need. The Peninsular Barracks at Winchester were converted into luxury housing. Crownhill Fort at Plymouth now incorporates holiday homes, managed by the Landmark Trust. Hillsborough Barracks at Sheffield experienced as great a metamorphosis as any – they became a supermarket, and the parade ground is used for parking cars and shopping trolleys. How many sergeants-major have turned over in their graves?

The Peninsular Barracks in Winchester have found a new lease of life as private housing. The former barracks, at one time the second largest depot in Britain, also house four military museums.

Further reading

The best architectural study of British barracks is:

Douet, James. *British Barracks 1600–1914: Their Architecture and Role in Society.* The Stationery Office, 1998.

For a study of the impact of barracks on the towns that host them, see:

Dietz, Peter (editor). *Garrison: Ten British Military Towns.* Brassey's Defence Publishers, 1986.

The following books have been found useful in giving a picture of life in barracks:

Cairnes, Captain W.E. *Social Life in the British Army.* John Long, 1900.
Cairnes, Captain W.E. *The Army from Within.* Sands & Co., 1901.
Costello, Con. *A Most Delightful Station: The British Army on the Curragh of Kildare, Ireland, 1855–1922.* The Collins Press, 1996.
Farwell, Byron. *For Queen and Country.* Allen Lane, 1981.
Grove, Doreen. *Berwick Barracks and Fortifications.* English Heritage, 1999.
Hardy, E.J. *Mr Thomas Atkins.* T. Fisher Unwin, 1900.
Henderson, Diana M. *Highland Soldier: A Social History of the Highland Regiments, 1820–1920.* John Donald, 1989.
Hewitson, Thomas. *A Soldier's Life: The Story of Newcastle and Its Barracks.* Tyne Bridge Publishing, 1999.
Muller, H.G. 'Marching on Their Stomachs: The Soldier's Food in the Nineteenth and Twentieth Centuries', in C. Anne Warson (editor), *Food for the Community.* Edinburgh University Press, 1993.
Nalson, David. *The Victorian Soldier.* Shire, 2000.
Neuburg, Victor. *Gone for a Soldier: A History of Life in the British Ranks from 1642.* Cassell, 1989.
Robertson, Sir William. *From Private to Field Marshal.* Constable, 1921.
Skelley, Alan Ramsay. *The Victorian Army at Home.* Croom Helm, 1977.
Spiers, Edward M. *The Army and Society, 1815–1914.* Longman, 1980.
Troyte, J.E.A. *Through the Ranks to a Commission.* Macmillan, 1881.
Trustram, Myrna. *Women of the Regiment: Marriage and the Victorian Army.* Cambridge University Press, 1984.
War Office. *Regulations for Supply, Transport and Barrack Services.* HMSO, 1908.
Wells, Captain John. *The Royal Navy: An Illustrated Social History, 1870–1982.* Wrens Park, 1999.
Williams, T. St John. *Judy O'Grady and the Colonel's Lady.* Brassey's Defence Publishers, 1988.
Wyndham, Horace. *Soldiers of the Queen.* Sands & Co., 1899.

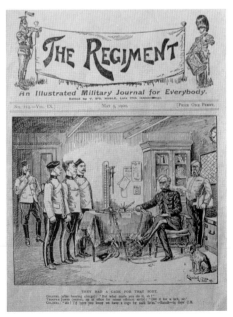

There were a number of military periodicals in the nineteenth century, but 'The Regiment' was unusual in that it was intended for circulation amongst the rank and file. In this cartoon, dating from 1900, a young cavalry trooper claims that he committed the offence with which he is charged 'for a lark'. To this, the colonel replies that there is a cage for such larks and sentences the man to fourteen days 'confined to barracks'. Such confinement, which was a common punishment for minor offences, would have been in the guardhouse.

Places to visit

Many historic barracks still form part of active military sites to which access by the general public is restricted. Some of these, however, are made accessible on Heritage Open Days. For information on access, and on other heritage issues, consult the Defence Estates website:
www.defence-estates.mod.uk

A number of military museums are based in former barracks or have displays relating to barrack life. A full list will be found on the Ministry of Defence website:
www.army.mod.uk/ceremonialandheritagemuseums

Another useful website is that of the Palmerston Forts Society:
www.argonet.co.uk/education/dmoore/index.htm

The following museums can be particularly recommended:

Aldershot Military Museum, Queen's Avenue, Aldershot, Hampshire GU11 2LG. Telephone: 01252 314598. Website: www.hants.gov.uk/museum/aldershot
Berwick-upon-Tweed Barracks, Northumberland. Telephone: 01289 304493.
Crownhill Fort, Crownhill Fort Road, Plymouth PL6 5BX. Telephone: 01752 793754. Website: www.crownhillfort.co.uk
Fort George, Ardersier, near Inverness, Inverness-shire IV2 7TD. Telephone: 01667 462777.
National Army Museum, Royal Hospital Road, Chelsea, London SW3 4HT. Telephone: 020 7730 0717. Website: www.national-army-museum.ac.uk

The building in London known as Horse Guards was constructed between 1750 and 1758 to designs by William Kent. It replaced a small guardhouse built on the site in 1649 and enlarged between 1663 and 1665 to accommodate both Horse and Foot Guards – the Sovereign's bodyguard. Until 1872 the central block housed the headquarters of the general staff. The side ranges provided barrack rooms and stabling. The adjacent Horse Guards Parade continues to be used for military ceremonial, such as Trooping the Colour.